WASHINGTON
D.C.

This book was devised and produced by
Multimedia Publications (UK) Ltd

Editor: Marilyn Inglis
Design: John Strange and Associates
Picture Research: Catherine Blackie, Tessa Paul, Picture Research – Washington D.C.
Production: Arnon Orbach

ISBN 0 8317 9309 0

First published in the United States of America 1984 by Gallery Books, an imprint of W. H. Smith Publishers Inc., 112 Madison Avenue, New York, NY 10016

Originated by D. S. Colour International Ltd, London.
Typeset by Flowery Typesetters Ltd.
Printed by Sagdos, Milan, Italy.

WASHINGTON D.C.

Michael Leech

Contents

The President's Legacy

Monumental Washington

Traveling the City

Hidden Neighborhoods

The Great Institutions

The Placid Backwaters

1 The President's Legacy

General George Washington was a remarkable man by any standards. Naturally he's best known as the first president of the fledgling United States of America, yet the man who gave his name to the capital of the country seems to hide behind his legend. A soldier and a statesman, he also knew a great deal about building – which must have stood him in good stead when he commissioned the architect of the future capital.

In fact, at the age of 17 George left the family home in Virginia (Wakefield in Westmoreland County) to take part in a survey of his half-brother's family's land across the Blue Ridge. The Fairfax estate was far from Westmoreland County but the young George enjoyed the challenge and, perhaps more important, the chance to explore what were then the frontier lands leading to the wide, untapped West.

When he was 20, George came into an inheritance he was to treasure for the rest of his life – Mount Vernon on the outskirts of the site of the new, and as yet undreamed of, city. From his land he surveyed the view that 40 years later – in 1792 – would be Washington.

Besides being a soldier (a colonel after his military initiative near Pittsburgh) Washington was also a farmer and a tobacco planter – and so inevitably a businessman too. His political career began when he protested against the intransigent and often downright foolish policies of colonial departments in far-off London, and his later leadership of the rebelling Continental forces is well-known. What is less well known is that as a surveyor and planner Washington was competent, practical and concerned; supremely able to oversee the development of the future city. Through the understanding and wisdom of George Washington, the city's first designer L'Enfant was selected – a man of considerable ability and vision.

Left The classic view of Washington in Spring – cherry blossom against a cold blue sky, and the monumental white marble architecture of the capital's monuments. The Jefferson Memorial features here.

This page right Buses and trains – Washington's road network is a busy one, and commuter trains run out above ground when they leave the city. Here famous monuments can be seen in close proximity to transport links – though carefully screened by trees.

Below Like many cities Washington's center is a mixture of styles – here a converted residence, now office accommodation, stands among modern blocks.

Facing page above An unlikely sight – a game of polo before the Washington Monument.

Below Imposing at any time of year, the winter months give a stark grandeur to the Capitol, glistening in the clear light. Hard to believe that that vast white dome is supported by a construction of cast iron!

Washington stands now as a memorial to many things, but its layout and basic look is as it was first envisaged almost two centuries ago. It is a neat, well-ordered city devoted to the business of government, as it has been since 1800. Although it has spread and grown both up and down with its own complement of underground developments, such as the splendid public transit system, it has not changed as much as most cities, and its center has not been overshadowed by gaunt skyscrapers. It is a city of wide boulevards leading to and from the White House and the Capitol, with many parks and open spaces to complement the huge number of government, ceremonial and commemorative structures that make up the Washington every visitor wants to see. Indeed, the so-called "Federal City" created by L'Enfant is still much in evidence, adding a sense of human scale to some of the larger, more characterless office buildings that have inevitably arrived to serve the many branches of government.

Besides being the seat of all federal government departments, Washington is the base for those who lobby officials and elected representatives, and is the home of many learned societies. Some, though not all, are closely attached to the government, but all benefit from this close proximity. Washington has little commercial development since its business still depends on the heartbeat of the country: its government.

Washington is a major tourist attraction and there is a well-equipped infrastructure to serve visitors. The city has a very good newspaper in *The Washington Post*, and much useful information can be gleaned from its columns on House of Representatives and Senate activities.

Right A colorful street market in the city.

Below The street is closed for a street-fair and everyone comes out to enjoy a promenade and open market stalls – and no traffic for a change!

Government and Business Center

Washington is indeed a fascinating place to visit at any time of year, and its principal occupation dictates that it should be well presented, clean and welcoming. It is not governed in the same manner as the other great cities of America – there is no elected council. Three commissioners are in charge, appointed to an executive board by the President and acting for the Congress. A population of less than one million resides within the boundaries of the city, but many more stream in to work every day from the sleepier suburbs of Virginia and Maryland, all within minutes of the city center. The District of Columbia, which distinguishes Washington's addresses from the state of the same name in the North West, is, to give it its full name, the Federal District of the United States, and it occupies exactly the same 70 squares miles as the capital itself.

Washington, then, is a working city as well as an ornamental one. It's a city of monuments, yet it's not a graveyard. It attracts the best in the arts and sciences to its doors, and has many cultural institutions

Left Solidarity and mutual concerns finding voice in music – Washington has long been the focus of many demonstrations and meetings and has seen triumphs as well as tragedies. From peace marches to pop concerts the city regularly attracts hundreds of thousands of visitors intent on special missions.

Below Luxurious shopping malls are the latest thing in the country and can be visited in Washington's smart and affluent suburbs.

offering information, entertainment and enlightenment to scholars, citizens and visitors. It's home for many thousands of Americans, black and white, and a temporary residence for innumerable foreign diplomats and businessmen. It's a great magnet for tourists and is a grand backdrop for many of America's greatest and most important events. It's a splendid symbol and a stirring reminder of the democratic roots and branches of America. And it's a lasting memorial to a plainspoken man who foresaw a new country. He was a man of "essential greatness" who was honored around the world on his death in 1799; a man who ensured that his country's capital was a beautiful as well as a useful place; a man of honor and simple beliefs – America's first leader, President George Washington.

Left An old-fashioned market hall on 0 Street. Looking like a cheerful red-painted toy, survivors of another age like this are now being carefully restored and utilized as part of the city's overall street scene.

Below This figure of white marble stands outside the Supreme Court with its pediment and Corinthian capitals. The building is open to visitors and often crowded on Mondays, when decisions are handed down.

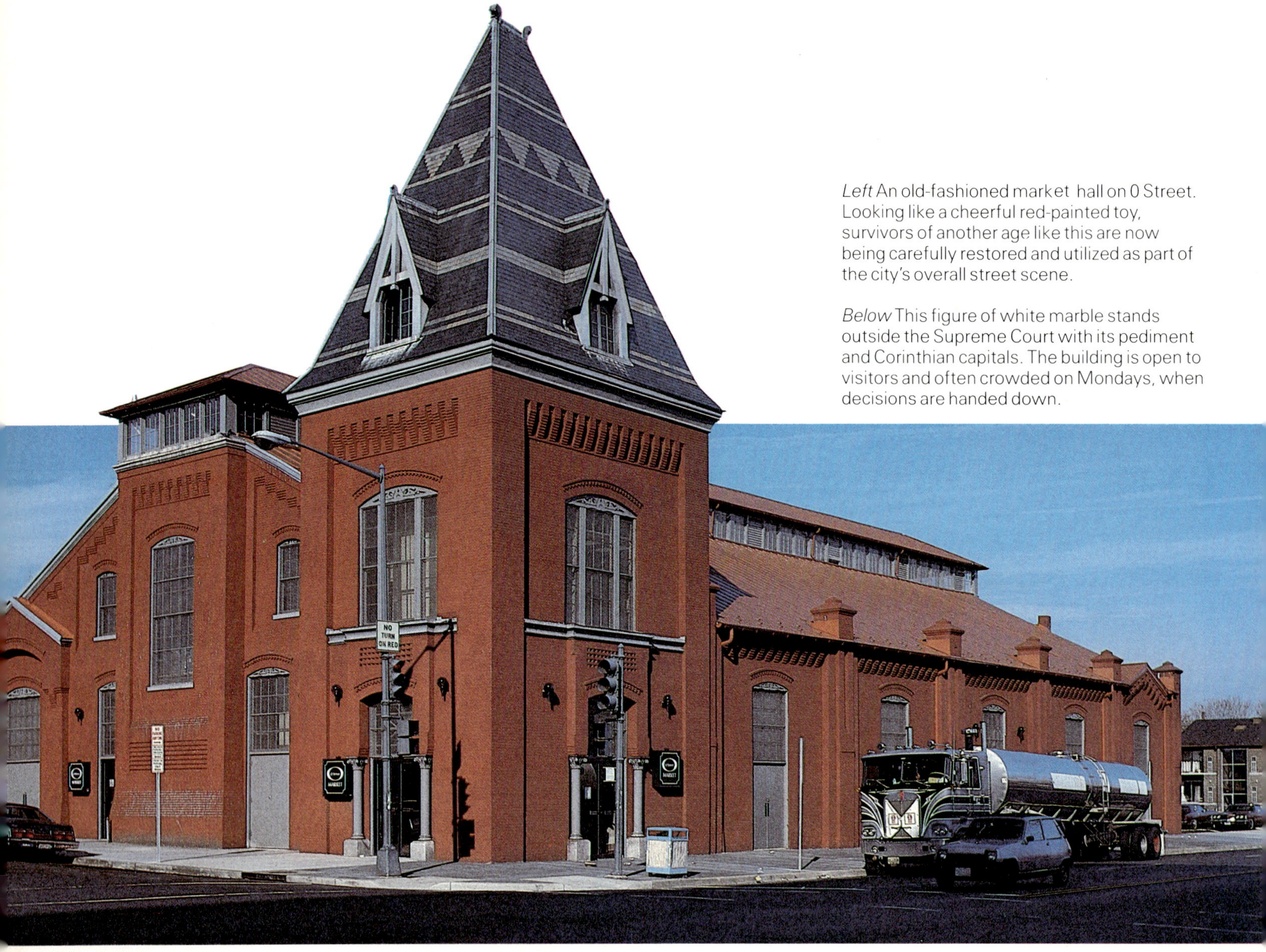

Left Redevelopment is common in parts of the city, where older buildings are giving way to the new. Here a mirrored building reflects and distorts an old neighborhood overshadowed by a vast office block.

Below The Hirshhorn Museum and Sculpture Garden is located in central Washington on the Mall. The circular building contains a collection of modern art, and the sculpture in the garden is grouped by a charming reflecting pool. It's managed by the Smithsonian.

2 Monumental Washington

Washington was built on history, but is also making history every day. Washington is the stage from which much of the world views America, from presidential appearances and great state occasions, to lively debates and the passing of laws. In Washington the visitor will find the true meaning and value of history – and perhaps come away with a new feeling for American civilization.

The first person to recognize the need for a city with a sense of history was George Washington – he chose the site of the future city and, with customary care, selected a magnificent area bordering the wide waters of the Potomac River in the states of Maryland and Virginia 70 miles square and in 1790 the District of Columbia took shape.

Washington was the first national capital in the world to be planned and laid out specifically as a capital. In 1791 a French officer, Major Pierre Charles L'Enfant, proposed plans for the capital city. He'd been noticed by General Washington after the War of Independence, having worked on Federal Hall in New York City where the General became the country's first president. The ideas the eager officer presented were truly grandiose in concept. He must have looked long and hard at Paris with its radiating boulevards, at Rome with its vistas and baroque grandeur, even at Christopher Wren's exciting plans for London after the Great Fire of 1666. He combined the best of these with ideas of his own and fashioned the capital of the USA. It was an inspired design; but alas the architect was unable to go ahead since Congress decided the scheme was too expensive. Later, long after the originator of the splendid scheme had died, Congress made an about face. In 1901, laws were passed to execute L'Enfant's abandoned designs, and the basic plan of the city is now much as he envisaged it, with triumphal avenues cutting across a grid of minor

Right George Washington and his generals at Yorktown 1781 are the subject of this James Peale painting. Courtesy of Christie's, London.

streets and utilizing major buildings as impressive focal points. If L'Enfant were alive today he would probably thrill at the view of the Capitol dome rising in great stateliness at the end of the grand ruled lines of Pennsylvania Avenue.

This page right A selection of memorabilia marking the Bicentennial in 1976.

Below Dunbarton Oaks, one of Washington's oldest buildings.

Facing page This portrait of George Washington painted by James Peale hangs in Washington University.

Second man on the scene was Andrew Elliott who laid out L'Enfant's scheme and worked on the White House, which began to rise in 1792, the Capitol building following soon after. Imagine it at the time – handsome buildings in open countryside, and, after the firing of 1814 by the invading British forces, blackened ruins for a while. Rebuilt and reinvigorated, Washington rose again and, although often in a dire position during the Civil War, it never fell again. It was, however, developing in an ugly, haphazard way, and after the collapse of the Confederate forces it was hardly a capital – more a straggle of buildings and muddy streets reminiscent of country lanes than a glorious city. Action was needed, and it's only a short time since Washington was transformed from a dull provincial town into its present status as the handsomest city in the USA.

Developing the City

From his elegant Mount Vernon home, George Washington watched the first attempts to make the city grow. He did not live to see Thomas Jefferson, the third president, inaugurated in the new capital, but from that time on the city itself witnessed some of the greatest and most stirring events in history, and since 1901 has provided a fitting and elegant backdrop. On television sets around the country Americans can see their first president's dream come true whenever there is a call for a great occasion of state. It's a delight to know that after L'Enfant's rebuff, his city is a reality. In 1909, the architect of the country's capital was brought back to Washington to rest forever in the National Cemetery at Arlington.

Monumental Washington

Washington is a city of remembrances, be they statues of Revolutionary heroes or, the most recent addition, a black marble wall by Maya Lin inscribed with 57 692 names to the memory of all those who died in Vietnam.

For many years it seemed Washington was indeed little more than a government center set in a cemetery. The many monuments scattered throughout the parks, squares and open spaces of the capital gave it a rather gloomy and depressing aspect – reverential but dull. That has changed vitally in the last 20 years, for Washington has made yet more history for itself by becoming a sophisticated place populated by lively, well-educated people who demand the many attractions of a major city.

Washington has always had libraries, museums and buildings; recently it has acquired theaters, art centers, shopping plazas, restaurants and cafés, and the city now buzzes with activity. Washington's image doesn't depend on everything being showily modern and new, however, like so many other cities. It was based firmly and decisively on the city's natural attractions.

When, in the early '60s, Washington decided it needed a cultural center, it set the John F. Kennedy Center by the river, in a bower of greenery, as carefully as a jewel. The same goes for the unabashedly modern addition to the National Gallery by I. M. Pei – the pale pink-white marble doesn't bow to the architecture of the city, be it Federal or Victorian, but its simplicity suits its background.

Washington has learned a very important lesson from its history as America's capital – the need for buildings to balance and complement each other, creating an unforgettable sense of what a capital city *ought* to be. And history in Washington adds that indefinable quality we call "atmosphere". The city has that in abundance. Far from being "bunk", as one eminent American industrialist suggested, history here gives America's capital city her charm, her elegance and her life.

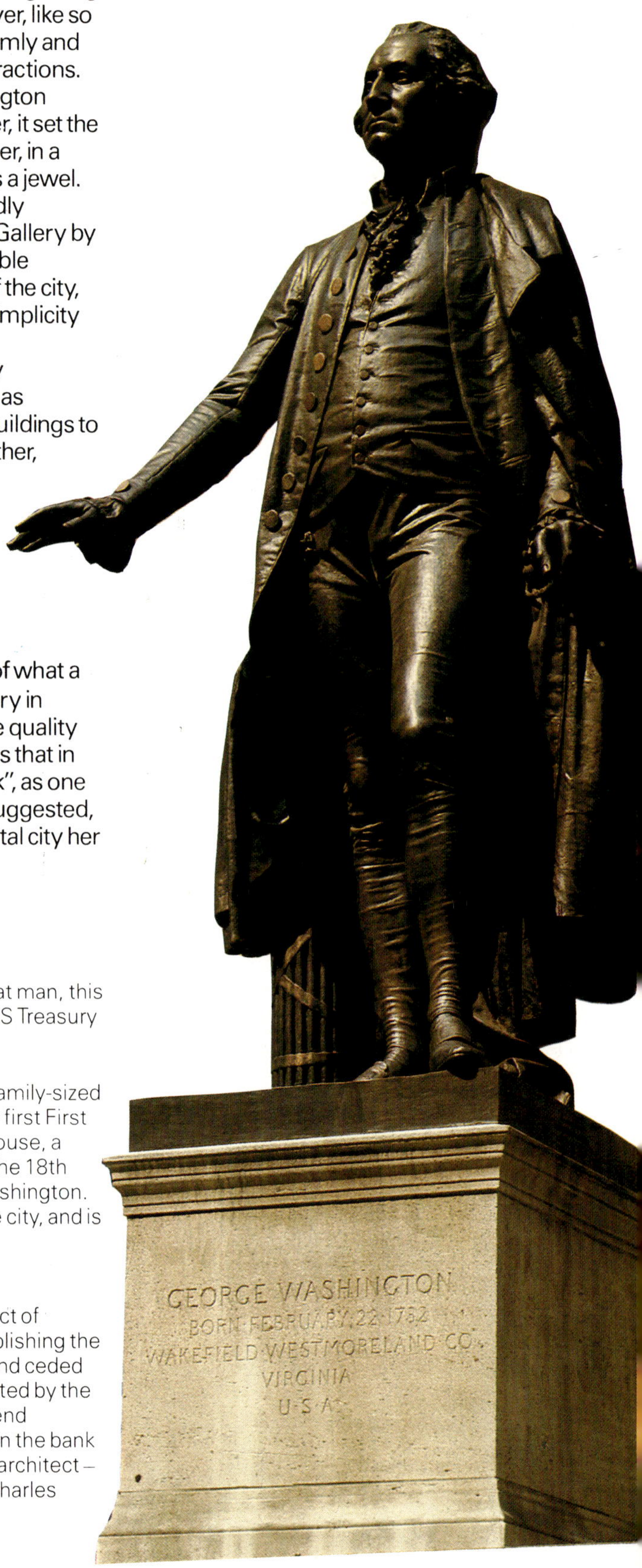

This page Another statue of the great man, this one in New York in front of the old US Treasury Building.

Facing page above The stately but family-sized mansion that was home to the very first First Family. Mount Vernon is a simple house, a country gentleman's residence of the 18th century, much loved by George Washington. It's in Virginia, a short drive from the city, and is well worth a visit.

Below This plaque in a park along Pennsylvania Avenue records the act of Congress of the years 1790-1 establishing the bounds of the new capital city on land ceded by Virginia and Maryland, and selected by the first President. Charmingly the legend establishes "the Permanent Seat on the bank of the Potowmac", and credits the architect – also anglicizing his name to Peter Charles L'Enfant.

PLAN
of the CITY intended for the
Permanent SEAT of the
Government of the UNITED STATES
Projected agreeable to the direction
of the PRESIDENT of the UNITED STATES,
pursuance of an ACT of CONGRESS passed the
sixteenth day of July, MDCCXC.
establishing the Permanent Seat
on the bank of the Potowmac
By Peter Charles L'Enfant.

3

Traveling the City

Washington was designed as a city with a sense of occasion. You get that feeling when you arrive at one of the city's airports (the National Airport handles domestic traffic; Dulles Airport, a 40-minute drive away in Virginia, all international flights; and there's a third at Baltimore, the Baltimore-Washington International) and drive into town. The most pleasant way to arrive is probably by Amtrak, which pulls into the monumental old station. Here are facilities for visiting the city – besides the ever-present taxis there is a coach service called a Tourmobile, and the station is a major center for the new and excellent subway system. If you have sturdy legs and comfortable shoes then try walking – you'll find yourself on Massachusetts Avenue as you emerge from the station's portico into its forecourt, and a short stroll down Delaware Avenue will bring you to the Capitol itself.

For those traveling by car, the city is served by major highways, Interstate 95 traversing the city from north to south, and Interstate 270 feeding in from the north-west. Coming from the shore side there's a tollbridge carrying Route 50. Exits into the city are well marked, but as in any major conurbation, parking in the middle of town is very difficult – it would be wiser to put your car into a garage and walk, or take the Tourmobile. (It stops at major points along a well-defined route taking in the principal sights of the city, and you can get on or off at leisure at no extra cost.)

For main places of interest it is best to refer to the list on page 70. There are also quite a few "secondary sights", and a stop at the Convention and Visitors Center will provide many suggestions once all the major monuments have been viewed. Here are the things you shouldn't miss along with their exact locations. Most of them are neatly and conveniently grouped around an axis between the Capitol building and the White House, with a preponderance lying south of the wide boulevard known to the nation as Pennsylvania Avenue, the route of presidents.

Late evening light lends a glow to the dome of the Capitol. The building stands on Capitol Hill, the nerve center of the nation. The bronze statue, by Thomas Crawford, is appropriately of Freedom, and stands almost 20 feet high.

The Central Sights

A good starting point is the White House, resplendent behind its iron fence and green lawns. You can view the ground floor state-rooms on a set tour, usually conducted on Tuesdays to Saturdays in the mornings only. You'll see the most famous and handsome of reception rooms from the biggest (the gold and white East Room) to the intriguingly oval-shaped Blue Room. Next to the executive mansion is the Treasury (to the left as you stand by the statue of Andrew Jackson astride his horse in Lafayette Square, the green space fronting the White House) and on the other side, the Old Executive Building. Next to the square is Blair House, a pleasing 19th-century building that is home for heads of state when they visit Washington. Walk down 17th Street beside the White House and enjoy the view from the elipse of the private gardens and great, bow-fronted south front of the famous residence.

Just ahead, piercing the sky, is the Washington Monument standing in a vast lawn, its height of 555 feet making it the most easily recognizable structure in the city. You can take an elevator to the top and enjoy a bird's-eye view of Washington, or simply soak up the surroundings from its base where the colorful flags marking the 50 states fly. Many famous names have appeared in public concerts given in summer in the Sylvan Theater, which lies to the south. From the monument you have a fine view to the dome of the Capitol along the wide space of the Mall, and in the other direction your eye will be carried along the length of the Reflecting Pool to the stern statue of Lincoln in his memorial Greek-style temple. If you walk this way through the beautiful Constitution Gardens, the wide stretch of the Potomac Tidal Basin is to your left and suddenly you'll see the lovely little circular temple that's the monument to that most cultured of presidents, Thomas Jefferson. In April it's submerged in a sea of pink blossom when the famous cherry trees light up the entire vista. It's a long walk but a rewarding one. If you have lots of time here you can take a break from walking, to ride a boat on the Potomac – Washington Boat Line Tours will ferry you along the river for a series of waterside views.

Below the Capitol's great white dome is General Grant's memorial. Along Constitution Avenue's straight ruled line at 1st Street is Senator Taft's belltower. Buildings ranked around Capitol Hill are mostly offices, but you'll also see the Supreme Court and the Library of Congress, and at the south-west corner is a

Facing page Even more impressive at night, when, like many Washington monuments, the facade is floodlit, the Capitol is the city's first and finest landmark. It owes its existence to a number of talents – Pierre L'Enfant selected the site, B.H. Latrobe began the rebuilding of the Capitol after it was burned by the British in 1814, Charles Bulfinch completed it and, 30 years later, T.V. Walter added the wings. In addition, a large section was rebuilt in 1861. Happily, even with all these disparate attentions, the Capitol emerges as a splendid statement in stone.

This page above The interior of the Capitol's dome. The great circular frieze under the windows depicts scenes from American history. Artist Constantino Brumidi began this epic task and, after he tragically died from a fall, other hands completed the work. The dome, made entirely of cast iron, covers the Rotunda where America's illustrious dead lie in state.

Below Like all Washington's monuments, the Capitol is usually thronged with tourists – here they take a chance to rest their feet.

Botanic Garden. You can tour the Capitol building all year round on a conducted visit, and without doubt the most spectacular room is the great circular Rotunda beneath the cast-iron dome. (From the base to the statue's head is a distance of almost 290 feet!) Look at the vast bronze doors named after the discoverer Christopher Columbus;

Facing page above The formal side of the White House as seen from Lafayette Square. The severe classical facade is relieved by lawns and flowerbeds. Beneath the pillared portico is the carriage entrance, eminently suited to the residence of the President of the United States.

Below Military bandsmen perform on the White House lawn.

This page above Second floor of the White House, where the First Family make their home in a number of rooms with beautiful views. Perhaps the most famous of these rooms is the President's office, the Oval Room.

Below A tour of the White House is a "must" for most tourists, who can follow a laid-out route through a series of grand rooms on the ground floor. The East Room, the Green Room, the Blue Room, the President's Ante-Chamber (also called the Red Room) and the impressive State Dining Room are seen on a typical tour.

at the paintings commemorating events in the life of the first president; the circular frieze depicting great moments of American history. Next to the Rotunda is Statuary Hall, where a whisper can be carried right across the room. Once part of the House of Representatives, it now houses statues from every state. The tour will usually allow a few minutes in either the Senate gallery or the gallery overlooking the House of Representatives. There's a warren of tunnels under the great mass of the building – including subways to nearby offices. If you want to stay longer in the

Left The keynote of L'Enfant's superb architectural plan – the 555-foot high marble needle that is the Washington Monument, seen here along the pillars of the Lincoln Memorial. Some idea of L'Enfant's design can be gained from the view of the Capitol in the distance along the length of the Reflecting Pool.

Right Millions of words have been written about President Abraham Lincoln, a mark of the greatness of this martyred president. A true native American, the 16th president of the USA is enshrined in his Memorial. His statue, seated on a vast marble throne, gazes out towards the Capitol – a moving sight, especially as dusk descends and the sculptured features seem to take on even more expression.

Below A simple building in approved classical style, the Lincoln Memorial is rightly accorded one of the most spectacular sites in Washington. The Gettysburg Address, tellingly brief, and an extract from his Second Inaugural Address (given just after the Civil War), are engraved on the walls of the building.

Capitol, apply to your senator or congressman for a special pass.

A walk along Pennsylvania Avenue gives a great sense of Washington's style as you pass such buildings as the National Gallery of Art, the National Archives, the offices of the FBI and the National Theater. Head north on 10th Street to a small yet important building – Ford's Theater, recently restored, where President Lincoln met his assassin. A couple of streets away is the National Portrait Gallery. Along Independence Avenue are the several buildings of the Smithsonian – it was started in 1852 when an eccentric Englishman named James Smithson willed money to a country he did not know in order to perpetuate his name! The original building is definitely an oddity, but since then there have been many additions and the Smithsonian now controls a host of areas, including the Washington Zoo. Its new National Air and Space Museum contains two marvellous exhibits – the

Facing page left In Arlington National Cemetery, the Custis-Lee mansion rises gracefully above the greensward. Below the pillared house is the grave of President John F. Kennedy, a simple slab beside which the Eternal Flame flickers.

Right The Tidal Basin and wide course of the Potomac River, viewed from the top of the Washington National Monument. Although illustrative of the distance one can see from the Monument, it also shows that Washington is actually better viewed from ground level – just as its designers intended. The white building is the Thomas Jefferson Memorial in West Potomac Park, dedicated in 1943 by President Roosevelt to mark the 200th anniversary of Jefferson's birth.

This page above One of the most cultivated of all American leaders was Thomas Jefferson, who served as president from 1801 to 1809. A remarkable, liberal-minded man possessed of many interests, he was the leading architect of his day. Jefferson's circular domed Memorial houses a bronze statue and also within the building, one of Washington's most charming, are four panels which bear passages from his writings.

Below The sparkle of Washington at night is shown in this picture, giving some idea of the grandeur of the city's public buildings when dramatically floodlit.

Wright brothers' airplane, and the plane in which Colonel Lindbergh made his historic trans-Atlantic flight.

Many Washington buildings are embassies and quite a few are offices to serve the billowing paperwork of the nation, indeed the world. One office that provides pieces of paper in the hundreds of millions is the Bureau of Printing and Engraving. Here, currency rolls off the printing machines at the rate of tens of millions of dollars a day! Stamps and government bonds are also printed at this office on 14th Street near the Tidal Basin, and you can watch this moving sea of money from a visitors' gallery.

Even though Washington is fairly compact the distances between places of interest can soon clock up the miles. Just these few places can take several days of sightseeing – so don't rush them. Washington has taken time to develop and the visitor should take time to enjoy the sights at a reasonable pace. After all, this is only the center of town!

Right Looking over Arlington National Cemetery towards the panorama of the city, with the Lincoln Memorial in the background. The crisp, clear days of early fall are a particularly pleasant time to visit Washington.

Left and below Presidential inaugurations are cause for celebration in Washington. These pictures are from President Reagan's inauguration.

AMERICAN
WORLD

Facing page Three busy airports serve the Washington area. Washington National handles domestic traffic while international flights take off and land at Dulles Airport, 24 miles away in Virginia. Nearby Baltimore, Maryland, also shares an airport with the capital.

This page left Besides the regular bus services provided by Metrobus, seen here, there are several specialist tour companies including the very useful and frequent Tourmobile which follows a set route around Washington's principal sights.

Below left and right Clean, efficient, swift and safe, the new – and still expanding – subway system is the fastest way to get around Washington. Entry and exit is by special card.

4 Hidden Neighborhoods

You have spent a few days in Washington, seeing and exploring the major sights – and found lots to occupy your time. There are also several very important attractions in the neighborhood of the city which shouldn't be missed. The pretty and altogether charming suburb of Georgetown is actually older than the city itself. You'll see evidence of this if you take Pennsylvania Avenue out to the north-west, past Washington Circle, and on into Georgetown. It has a character all its own, yet is indubitably part of Washington and provides many of Washington's citizens with handsome homes – for Georgetown belongs to that period known as Federal, a style close on 200 years old, with its roots in English Georgian architecture. Well-proportioned houses of soft red brick, with small-paned windows and slate roofs, often with handsome doorways – these are the houses that modern home builders like to emulate.

Georgetown oozes atmosphere, and although it's as neat as a film-set, there aren't actually any major buildings to visit. True, there are museums such as Dumbarton Oaks on R Street NW with gardens as well; and Georgetown University is also to be found here, but by and large the area is simply a stroller's paradise. Lots of leafery draped over the brick sidewalks and quiet streets to explore – it's a bit like Boston's Beacon Hill, but without the steep inclines! You'll find lots of camera subjects and delightful backgrounds here.

Georgetown's shops are renowned – small, selective, and usually expensive, they offer all sorts of unusual things. You'll find them – along with a plethora of galleries, restaurants, bars and clubs – along Wisconsin Avenue or M Street NW. These two thoroughfares mark the center of Georgetown, so take time to explore both. You can walk along the towpath of the

Left Neat, cared for terraced houses mark several of Washington's more affluent suburbs, such as the picturesque and appealing Georgetown.

Chesapeake and Ohio Canal, and another way to enter the area is across the meandering Rock Creek Park over the curious Buffalo Bridge – stone statues of the beasts mark each end. A walk along Rock Creek as it flows below the city's buildings offers a chance to get away from the maze of brick and stone for a while, and you can stay in this narrow green oasis bordering the creek until it debouches into the Potomac River.

Memorials and Mount Vernon

Washington's embassies are scattered throughout the city, but you can view a group of them in the neighbourhood known as Embassy Row. They are massed north of Massachusetts Avenue along 23rd Street, and many are architecturally notable. Small plaques and flags tell you which country owns which house, though one or two, such as the British Embassy, are so big you can't miss them. You may also be lucky enough to see the Vice President riding past – the official residence is in the gardens of the Naval Observatory nearby. (The office of the VP is in the Old Executive Building, next door to the White House.)

Across the river, viewed from the hillock on which the Washington Monument

Above Small, smart and inevitably expensive shops line the cobbled streets of Georgetown, a suburb that's older than Washington itself.

Left Perhaps one day such potentially attractive dwellings as this small terrace will be restored. Central Washington contains a large pocket of poverty, contrasting uncomfortably with the areas housing privileged diplomats and officials.

stands, you can see a mansion beyond the length of the Arlington Memorial Bridge. It's a striking Greek-style pillared building standing on a hill. This is the Custis-Lee house and is famous as the home of Colonel Robert E. Lee. Arlington House – its other name – is open every day and can be visited when you make your obligatory visit to Arlington Cemetery – the Tomb of the Unknown Soldier mustn't be missed. In addition, there is the eternal flame fluttering over the grave of assassinated President John F. Kennedy (his brother is also buried here) and the Amphitheater. The Arlington National Cemetery is a lovely spot, despite its sad atmosphere, and is open from spring to fall.

Be sure to visit Mount Vernon, to view the panorama Washington himself knew so

This page above From the air, Washington's affluent suburbs look like those of any major American city with their maze of good-sized houses and swimming pools. In the capital itself there's more parkland than usual, for this city abounds with green spaces and pleasant lakes, and the highways are edged by trees.

Right Along the grand sweep of Embassy Row, fronted with flowerbeds and waiting limousines, are the palatial town houses that serve as embassies for the many foreign countries represented in Washington. Each is officially designated as a small piece of the country it represents. If you've a good knowledge of flags, you'll recognize the individual countries at once from the national banner flying outside each house.

Facing page Washington has its share of older, run-down apartments, much sought after by the hard-up students of this university town.

well. You won't be disappointed – this country house of a well-to-do gentleman of 200 years ago is replete with charm and nostalgic atmosphere, and there is much to see. Allow plenty of time, although a package tour to Mount Vernon is available by taking the Washington Boat Line trip offered every day from spring to early fall. There are extensive gardens, and you can walk around the house which is beautifully furnished as the Washington family themselves would have known it. It's certainly not a grand house, and the spirit of the first of the First Families comes across as you walk from room to room or around the outbuildings and grounds. Both the President and his wife, Martha, are buried here. Although a fair distance from Washington (about 40 minutes' drive, so allow several hours for the visit) this is one place you should certainly not miss, and it can be combined with a look at the suburbs of Virginia, south of the city.

On the way to Mount Vernon is the handsome town of Alexandria, a place packed full of historical reminders, from the Tomb of the Unknown Soldier of the Revolutionary War (at the Old Presbyterian Meeting House), to a number of delightful 18th-century houses. If you're going by car and have plenty of time, look at the extensive suburbs here and in Maryland, for this is where many Washingtonians actually live, commuting into the District on working days. It's therefore wise to avoid rush hours!

A Feast of Houses

Washington has so many major attractions that the less well-known, yet often very fascinating places, may sadly go unvisited. Besides the many museums, there are unusual houses and places to see everywhere. Suggestions for a series of alternative visits are listed below.

Decatur House, neighboring the White House, can be visited at its address on Lafayette Square – it's a town house of considerable elegance designed by architect Benjamin Latrobe. Dumbarton House in Georgetown is even older, and is the office of the Colonial Dames of America. It offers the chance to see a fine collection of antique furniture, porcelain and silver. You can see similar displays at the Daughters of the American Revolution Memorial Continental Hall. The Zoological Park is in Rock Creek Park and has a varied collection of animals on show. Also in the park is an operating mill, and there's a planetarium in the Nature Center. A variety of boat trips are on offer – but why not take a swan boat out on the Tidal Basin, or row yourself in a

NGINE
D. C.
RE D

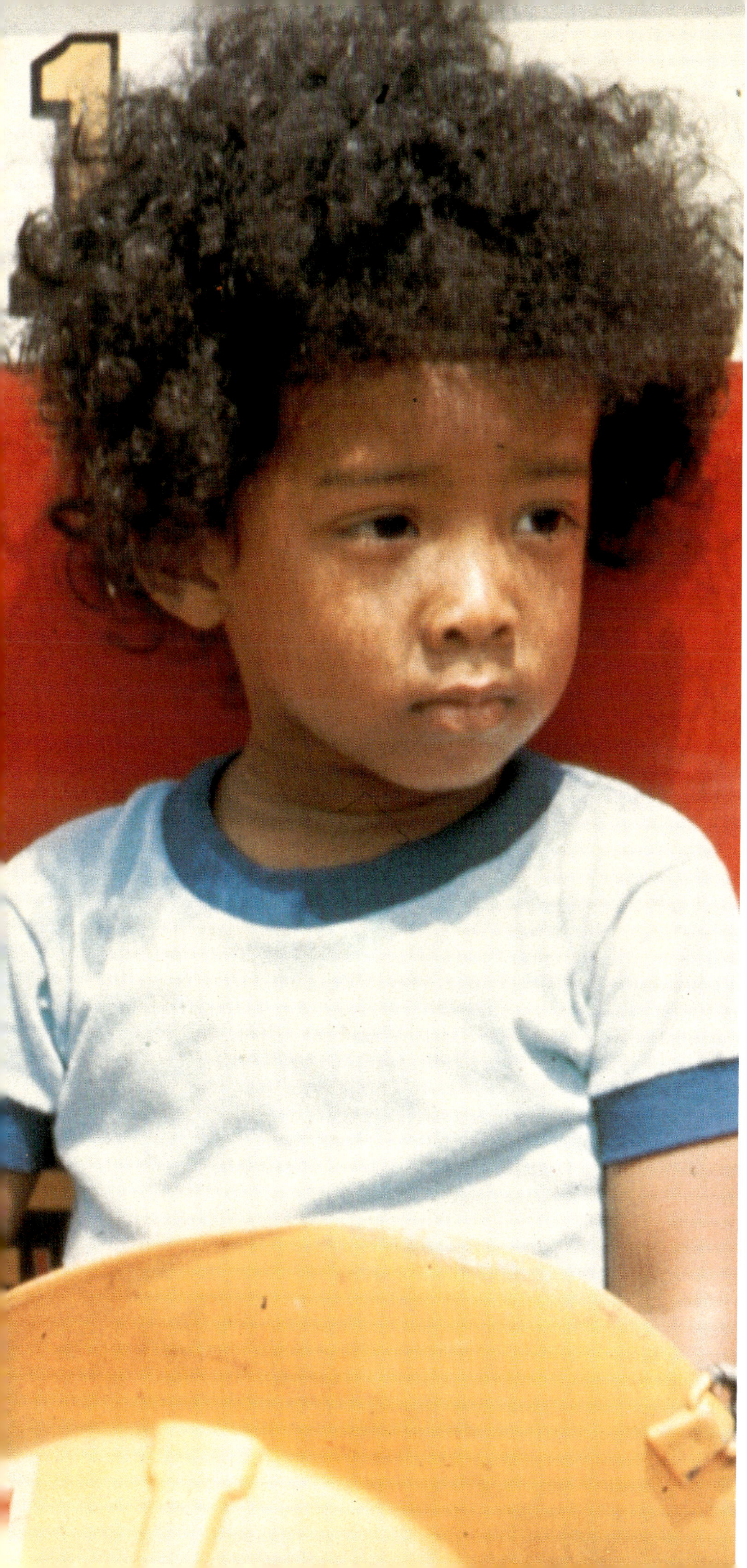

rented craft for a quiet hour or two? You can take a short cruise on *The Spirit of '76* to Georgetown and back. Interested in Woodrow Wilson? You can visit the house where he lived and died on S Street. Try a free tour at the FBI building. For the sports-minded there are several horse tracks close by, and harness racing can be viewed too. Ice hockey, football, basketball and even polo are played in various locations, and just to show that Washington really is one of the most unusual of all American cities, you can even see British cricket played, usually on summer weekends, in West Potomac Park!

Traveling the city

Vast numbers of Washington's workers don't actually live within its boundaries, but cross its borders every day to their homes in Virginia and Maryland, and even as far as West Virginia. They trek in every day by car and bus (Metrobus provides a good service at a flat fare) and the excellent subway system – which is, perhaps, the best way to travel, while others are struggling along in traffic in their Gray Line and Diamond tourbuses!

The subway system, which is fairly new, spanking clean and well-designed, is still being extended, but already you can zoom along on smooth, comfortable, rubber-wheeled carriages to such places as the Pentagon, the RFK Stadium, the central railroad station and the domestic airport. It operates on a flat-fare system within zones, using special cards which have to be fed into machines – children particularly enjoy doing this! Open from early morning until midnight (it shuts down earlier on Sundays) it's a great way to get about, is safe, swift, and inexpensive too. Metro stations are marked with signs painted brown bearing a large capital M, and there are extensions by bus at the end of the line.

There is a large black population in the center of the city, and the flavor of the shops, restaurants and clubs is decidedly different from the mainly white areas, such as Georgetown. In some of the poorer

Left Washington's summers are often unbearably humid. This tot standing by the big red fire engine looks like she's hoping some nice fireman will come along and open a hydrant so she can cool off – or perhaps, since she's holding a fireman's hat, she's already got the whole thing arranged!

areas, attempts are being made at urban renewal. It's interesting that a city that entertains so many of the well-off in government circles should also have an element of poverty. But it's as well to remember that Washington is almost in the South and, as with New York, there has long been migration from the poor rural areas into the more affluent northern cities. The government offices in town do provide a good deal of work, but since there are not many other employment possibilities quite a few of Washington's poor remain unemployed.

It is interesting too that residents of the District of Columbia are not allowed a right many of us assume is basic to all Americans – the right to vote. They cannot vote for local councils (there is a board of commissioners, no mayor or councillors) and so they have no say in the governing of the city. They also have no representation in Congress, and the city has no senators of its own. And it was a recent innovation – the adoption of the 23rd Amendment to the Constitution in 1963 – that finally allowed Washingtonians to vote for President and Vice President.

Besides its population of government employees the city also has a large complement of students since there are five universities within the District from the venerable Georgetown (1791) to the relative newcomer, the American University founded in 1893. The others are George Washington, Howard and Catholic universities. They help to add a youthful zest to the city, and to experience that just tour around the many clubs along The Strip in Georgetown on almost any evening. This is *the* place for those who like nightlife, and there are several clubs that provide good jazz, blues, rock, '20s music, bluegrass and soul, and even sentimental Irish folksongs! The crowd is usually young and full of energy, and you'll be able to see a side of Washington that is anything but sedate.

Left Washington has followed a fashion set in New York, London and San Francisco – street art on a massive scale. Here, an artist touches up Marilyn Monroe's sweeping eyelashes as she stares from a wall, pouting those unmistakable lips.

Facing page right Washington has many neighborhoods populated by ethnic minorities and every so often there's a procession with lots of lively music and fun. Everyone loves a parade and Washington was just made for 'em!

Left Summer in the city means sidewalk art shows and a chance to socialize with the neighbors – and maybe get to know your local cop, too.

Below The Fourth of July in Washington – even though it's hot it's time to celebrate. In the capital city Independence Day is greeted with all the usual razzmatazz.

This page right Is this seller scratching her head at the sheer variety of fresh fruits and vegetables, or just wondering where to put everything? Washington's open markets ensure a daily supply of fresh produce, and there are truck farms only a short distance away from the capitol.

Below Nearby Chesapeake Bay supplies the city with splendid fresh fish, shrimps and crab. These form the basis of many a local delicacy, and Washington's restaurants often serve up Southern recipes for fish, too. These shoppers on Maine Avenue are looking for seafood to take home.

Facing page above Watch out – sidewalk speeders. These kids may live in a seedy suburb now, but perhaps one day they'll help make Washington a better and fairer place to live. Until then – look out for those wheels!

Below Besides an assortment of major supermarkets, Washington has many small corner stores that stay open at all hours.

THE MARKET
THE AMERICAN CAFE
DON'T COOK TONIGHT
CARRY-HOME
ENTREES
SANDWICHES
CROISSANTS
SALADS
CHEESE
DESSERTS
THE AMERICAN CAFE
MARKET
THE AMERICAN CAFE

5

The Great Institutions

The great institutions of the city are something Americans are familiar with yet perhaps don't really understand. Of course the business of legislating and governing the country through its elected representatives goes on at the Capitol. But what is done at the National Archives, the Library of Congress, or even the Supreme Court? All these institutions, and many more, can be visited and will often provide a tourist with a valuable insight into how America works.

Take the Supreme Court, a magnificent building with its numerous Corinthian pillars and ornate pediment in the Roman style, as borrowed from Greece. You'll find it to the east of the Capitol and, when sitting (October to June), the actual processes can be observed from 10 a.m. to 5 p.m. Making the laws of the United States is a solemn process, and due pageantry is observed when the Supreme Court Justices make their appearance, announced by a bailiff. There is a public gallery from which to view the proceedings of the highest court of the US judicial system and the third branch of the government, which is proudly independent under its Chief Justice and his eight Associate Justices. It pays to read up on some of the history of the Court before attending a session so you can follow procedures, but note that Decision Day is usually Monday, when things may well be more exciting than usual. During the summer recess there are tours of the building.

The Library of Congress is next to the Court, by the south-east corner of the Capitol's grounds. It's a huge building containing manuscripts, documents, pamphlets, newspapers, magazines, sheet music, prints, photographs, films, tapes – and, of course, books. For over a century the Library has been the office of repository for all copyrighted items, and the collection is added to at the staggering rate of two million new items a year. Amazing, when you learn that it all began in 1800, in one

Right The tragic event that took place in this ornate little building reverberated around the world. At Ford's Theater, on April 14, 1865, just four days after the bloody Civil War ended, President Lincoln was shot by John Wilkes Booth.

room! This great research library contains the libraries and documents of many presidents and leading figures. All carefully preserved, from rare musical instruments to modern microfilm, the vast collection is housed in two buildings. There's the main mid-Victorian structure with dome and carvings, and in contrast the new sharply-modern extension, the James Madison Memorial Building. In the auditorium of the Library of Congress there's a continuous program of lectures and concerts, and poetry is a recipient of special sponsorship. There's a regular hourly tour of the building, or it's possible to browse round alone, admiring the treasures for yourself. It's open weekdays from 9 a.m. to 4 p.m.

Close to the Library of Congress is the Folger Shakespeare Library, set up by Henry Clay and Emily Folger as a study-place to "help make the US a center for literary study and progress". The couple gathered together an enormous collection of works concerned with William Shakespeare, and eventually founded the Library as a research center for Anglo-American relationships and concerns of Shakespeare's time and the two centuries following.

Exhibitions and Art

In the National Archives it's possible to see some of America's great moments enshrined – the Declaration of Independence, the Constitution, and the Bill of Rights, set out in the exhibition hall. This is much more than a museum, however – it houses and cares for all the federal government records, and also acts as a publisher, being responsible for the publication of statutes and the

Facing page above Another view of the Capitol.

Below The inventive cameraman or woman can find all sorts of unconventional angles from which to photograph the enormous Capitol with its grandiose facades.

This page above Soldiers parade in full dress at a ceremony at the Tomb of the Unknowns in Arlington National Cemetery. The cemetery is actually in Fort Myer, Virginia, but is administered by the Department of the Army.

EQUAL JUSTICE UNDER LAW

papers of presidents, as well as producing a daily news sheet, the *Federal Register*. This center of records can be visited every day.

The National Gallery is the largest marble building anywhere, and was built in 1941, having been established by an Act of Congress. Andrew Mellon endowed it and donated its nucleus of great paintings; since then other collections have given works so that it now holds one of the most important collections of art in the world. It consists of a rotunda with galleries radiating outwards, and hundreds of paintings are displayed in

Facing page Spectacular in its architecture, the temple-like white marble structure is the Supreme Court, standing next to the Library of Congress and across the street from the Capitol. The Court is the highest level in the judicial process and covers all cases that arise under the Constitution.

This page above This ornate ceiling shows how fashions in official architecture change – it's in the highly approved Italian Renaissance-cum-Baroque revival style of the latter part of the 19th century. It covers the Reading Room of the Library of Congress – a building that was three times a victim of fire in the early years of the Republic. By contrast, the extension of the library is uncompromisingly modern.

Left This circular room is a place of silent activity – it's the main Reading Room of the Library of Congress. A visitors' gallery runs around the vast room at second-floor level. This is one of the great libraries of the world, housing scores of millions of literary items – Jefferson's own draft of the Declaration of Independence is on permanent show.

near-ideal conditions. Many are masterpieces of their period, but there are no contemporary works. The nearest you will come to that is the startlingly modern annex, built in 1978, and known as the East Building. On Sundays, in the fall to spring season, free concerts take place at the Gallery.

To see indigenous art go to the National Collection of American Art which is combined with the National Portrait Gallery. Here you will find special shows as well as permanent displays. It's part of the Smithsonian Institution and is one of those eccentric establishments that endear the visitor. It is famed for a number of reasons. It took ten years for Congress to decide that it had the right to take James Smithson's £100,000 bequest, but eventually it did and the Institution's doors opened in 1846; it was the earliest foundation for the pursuit of science in the young Republic.

Since those early days the organization has pursued research in all scientific areas, and the fruits of its labors can be seen in its enormous collections. In the Museum of American History, the Museum of Natural History, the Freer Gallery of Art, the Hirshhorn Museum and Sculpture Garden (modern and late 19th century art) and the National Air and Space Museum you can gain an idea of the scope of the organization. Like the National Geographic Society (also based in Washington, and founded to assist and encourage research

This page above The symbol of the USA, the proud bald eagle, can be seen in stone on many public buildings in Washington. This one has his perch at the Federal Reserve Building.

Right Take a tour of the Capitol and you may see a debate if Congress is in session. Free tours include a brief visit to the public gallery – an ideal opportunity to view the "living" side of Washington, its business of governing the country.

Facing page above In its isolated setting on the Potomac River, the John F. Kennedy Center for the Performing Arts offers wide views from its terraces and roof-garden. Within, there are handsome auditoria for theater, opera, music and film presentations. This building, by Edward Durell Stone (who also built the National Geographical Society's headquarters), injected new life into the city.

Below Washington's embassies are a sight in themselves. Many are large town houses, but some, such as the Venezuelan Embassy shown here, are mansions surrounded by park-like gardens. The French and British embassies are also particularly impressive.

and exploration in the field of geography) the Smithsonian publishes a monthly illustrated magazine called *Smithsonian.* The range of exhibits is vast, everything from the Hope Diamond to exotic and ornate Victorian artefacts at the Renwick Gallery.

There are many other special collections on show in the city – from African and Russian art to textiles and rugs. Pick up a guide to Washington from a news stand for up-to-date information on the many temporary exhibitions that can be seen. Washington's cultural life is varied and you'll also find a number of private galleries offering shows; indeed, theater foyers often house small exhibitions. Most museums have a souvenir shop, so you can take back mementoes of your visit. A stroll through the campuses of the universities will also reveal special events, lectures, performances and demonstrations.

The activities of the city reflect the interests of the population, which tends to be well-educated and aware. It's not surprising that institutions of all sizes and scope flourish here, and in addition to the traditional sights, the keen visitor will find much to stimulate and enliven the mind in this cosmopolitan environment.

This page above A severe, functional design is now the accepted norm in Washington. The East Building of the National Gallery of Art is by I.M. Pei and was opened in 1978. Since that time, the pale pink-white marble structure has become an attraction in its own right.

Right Washingtonians maintain a genuine affection for this rather odd, turreted building on the Mall. It is the Smithsonian Institution, built of brick in 1852, with funds donated by an Englishman who had never been to North America! The institution now administers many of Washington's museums and galleries.

Facing page right Museums in Washington are spacious places, and the Museum of Natural History is no exception. If you tire of looking at the exhibits, you can rest a while in cool high-ceilinged foyers where fountains splash among exotic plants.

6

The Placid Backwaters

You will have quickly gathered, from the descriptions of Washington and the photographs of the city and its surroundings, that this is a place gloriously endowed with parks and green spaces. In that respect, Washington is a model city.

Many of the plants were transported long distances to grace the capital's gardens – the famous cherry trees, a foam of bloom in early spring, are from Japan. At the Pan American Union near Constitution Avenue at 17th Street are trees and plants from the 21 South and Central American republics represented by the Organization of American States. Many of the embassies feature their country's plants in their gardens, and in addition trees from all over the USA can be seen in the numerous green pockets of the city. Look, for example, for the giant tulip trees in Rawlins Park, and the old-style shade trees in Rock Creek Park.

Throughout the year gardeners strive to keep Washington alive and lovely with flowers – even in winter, when the city gains a special magic from the touch of snow and frost. Somehow, even though the great buildings look splendid framed in greenery, they look even more impressive in the whiteness of winter. It's then that you get an impression of the stateliness of the White House, with the rose garden wreathed in snow, or of the grandeur of the monuments, removed from the everyday and lit in the odd, eerie light that winter gives. So if you can't get to Washington in spring or summer, you may find a quite different yet equally alluring city in the colder months. Washington has fall colors, though they're much more muted than the famous brilliance of the New England woods.

Washington is also different by night. For then many of the great buildings are floodlit, adding a striking dimension. It's at this time that the numerous fountains come into their own. Lovely by day, they take on a fairy-tale quality at night, and challenge the specialist camera bug to catch their beauty.

Right A dramatic gull's-eye view of the Potomac River. Washington utilizes its wide, winding river and one's never too far away from it when driving around the capital's highways.

When all these appealing features are put together they make the capital something the pragmatic and prudent George Washington probably never dreamed of – a romantic spot that holds the eye with its planned yet unpretentious expressions of sheer beauty. The place seems a more apt honeymoon spot than Niagara Falls! Cherry blossom makes a natural background for newly-weds, so also do a hundred other Washington scenic sights.

Washington shopping is exciting and varied. An area north-east of the White House, cut by New York Avenue, is the main shopping area, with department and speciality stores. Small shops are found around Wisconsin Avenue in Georgetown, and further out at such centers as White Flint in Maryland, and there's one of the fancy 'gallerie' or modern arcades, called the Mazza complex. At the confluence of Western and Wisconsin Avenues are more department stores, and the new Watergate Complex has unusual and expensive smaller shops and boutiques. There are also new malls in the suburbs.

Facing page above The Potomac is a commercial river – it handles freight shipping, as well as a great many small boats and pleasure craft, sometimes including a presidential yacht! As this picture shows, it is bordered with lawns and trees in many places and landscaped along its route through the city.

This page above Georgetown's canal is a surprise to visitors – it offers a pleasant and almost rural walk through this leafy suburb. Its full title is the Chesapeake and Ohio Canal and you join it at Thomas Jefferson Street and 30th Street, where the neatly-bricked towpath begins. This traffic free way is a favorite with joggers!

Left The Chesapeake Canal has several bridges and locks along its length. Take a lazy, slow journey on one of the old narrow-boats.

Facing page left Georgetown is not only a smart suburb and Washington's best residential area; it also possesses many night clubs, bars and restaurants so that on any evening along M Street or Wisconsin Avenue you'll find crowds bent on having a good time. Here musicians warm up for a session in a small club.

Below Members of Washington's large black population know how to have an enjoyable evening at one of the many bars and nightspots of the city.

This page right As you would expect, Washington has a lion's share of good restaurants – for with many government officials dining is part of life and work. Here, in a subdued setting, diners lunch at the Jockey Club.

Below Like so many cities, Washington has developed an 'indoor/outdoor life', with covered streets and shopping plazas, often gaily decorated and bedecked with plants. Here a lunchtime crowd chats over drinks and food at the Old Post Office Building, a recent development of shops and walkways.

Food and Fame

Restaurants abound, and the variety suits all sizes of purse and appetite. No one should miss such local delicacies as Maryland crab or the southern taste of Virginia dishes. Ethnic restaurants are everywhere, from Italian to Vietnamese, Japanese to Greek. If you are dining in one of the many eating-places close to the White House or the Capitol you may well be able to celebrity-spot when the Senate is in session.

It's possible to see stars at the Kennedy Center – both from the stage *and* the heavens, if you dine in the grand Roof Terrace Restaurant. The Center helped to turn the tables on all those folk who

Left Although it's not in the center, you won't want to miss the chance to see Gothic-styled Washington Cathedral.

Above A principal Roman Catholic church is the grandiose Shrine of the Immaculate Conception. Its belltower and gleaming dome make it a Washington landmark.

Right A famous small church – St. John's on 16th Street just up from the White House – is famous as 'the church of the Presidents'. Built in the early part of the 19th century, to designs of Benjamin Latrobe, it has received almost every president into its handsome classical interior.

PAR

monotonously complained, less than two decades ago, that Washington was dreary after dark. Bringing in close to 8 000 spectators to its three major performance spaces every evening, the Center created a need for restaurants and bars in the heart of the city and, as a by-product, rendered dark streets much safer by simply ensuring that there were many people using them as they came and went to the show. Alas, the work of Edward Durell Stone hasn't been standing up to the test of time; cracks have

Below Before spring has brought out the first leaves there is a magical short period, in early April, when the hundreds of Japanese cherry trees blossom. Then this part of the Tidal Basin is a wash of pink bloom for the annual Cherry Blossom Festival. The handsome circular building is the Jefferson Memorial, its pale stone is a perfect foil for the delicate pink flowers, a photographer's dream.

Above Washington has many charming small buildings along its tree-lined streets, many of them aping classical structures, although they were built in the early part of this century. This town house is home of the Cosmos Club.

Left Made famous from the famous scandal of the Nixon years, Watergate is actually an impressive structure of apartments and offices. The area it stands in is called Foggy Bottom due to the low-lying land and the prevalent river mists.

Left Winter in Washington is a special time. Even in the chilly season you might consider a visit for it is uncrowded and may well offer unusual scenes in the snow. Here skaters enjoy the ice near the Smithsonian Institution.

Below It's not uncommon to see a group of people being entertained or actually entertaining, at the Capitol. Here, on one of the broad terraces, a college band in Highland dress rehearses before a concert.

Right In summer, Washington takes on an almost Parisian air with small sidewalk cafés offering drinks and snacks, as well as full meals, outdoors. These diners are eating outside at La Brasserie, near Capitol Hill.

La Brasserie

UNCOMMON
VALOR
WAS A COMMON
VIRTUE

already started to appear in certain parts of the Grand Foyer. But the Grand Foyer is still undoubtedly one of Washington's sights. It's 630 feet long and very high, and is the place to promenade during intermissions. There's also a Riverside Terrace with fine views, and don't miss a trip up to the roof with its gardens and places to sit and relax. Regular tours take place on weekday mornings and offer fascinating backstage glimpses of the theaters, opera house and concert hall.

There are also several small "off-Broadway" type theaters in Washington for live shows, and some cabaret too. Check for outdoor musical and dramatic events during the warm summer months. These can be memorable, set against the floodlit backdrop of the capital city. And, like so many of Washington's attractions and tours, many events are free. Washington has its complement of street entertainers, performing, as such artists have done ever since cities began, with a lot of sass and a hat placed hopefully around.

By now you will have gained some knowledge of the capital city of the United States, and captured a *feeling* for it. On the surface Washington's a formal city, dedicated to a very important yet not easily pictured activity, the business of government. Under that smooth, unruffled surface it's a hive of diverse and growing activity. In a way Washington is *every* American's hometown, and you can get that feeling of propriety from each tour bus or group of visitors. Everyone feels they know something about Washington before they arrive; with luck they will carry away a more detailed picture, and one that will give pleasure as they reflect upon it.

The city is a focal point of the country, not just at election time but whenever the business of governing the country is going on. The picture's an elegant and well-designed one, thanks to the brilliance of Pierre L'Enfant and his assistants, who created a series of marvelous scenes. Over the years Washington has become a desirable place to live – you can no longer say of it, as the cliché-lovers say of New York, "nice to visit, wouldn't want to live there". Washington has become a very attractive place, with a genuine sense of human scale in its make-up. If L'Enfant and George Washington could come back and see the city they envisaged, one feels that they would not be unhappy with the result. Washington DC is a city of the present, *and* the future, but it is just as proud and concerned about its past.

Left Washington is a city of memorials, like this stirring monument dedicated to the struggle for Iwo Jima in the Pacific in 1945.

Major Attractions

1 The White House Open daily except Mondays and Sundays. Tours from 10 am to 12. Set route through eight staterooms on ground floor.

2 The Capitol There are regular 45 minute tours, and afterwards you can look round alone. Tours go from 9 am until 3.45 pm and are free. It's possible also to take a ride on the special subway which joins the Capitol to its neighboring office buildings.

3 The Supreme Court Regular tours, although the court is open from October to June only for sessions. Weekdays (except Sats) 9 am to 5 pm.

4 The Library of Congress Regular hourly tours of this building and its new extension. You can also explore on your own, and the Visitors' Gallery should not be missed for its view into the massive reading Room. Open weekdays (except Sats) 9 am to 4 pm.

5 The Folger Library Free entry and wander on your own to see the vast collection of Shakespeareana put together by Henry Clay Folger and his wife. This fascinating collection is also a research institution.

6 Arlington National Cemetery Tomb of the Unknowns, with the bodies of soldiers who fell in World Wars 1 and 2 and the Korean War. Graves of President Kennedy and Senator Robert Kennedy. Also visit Arlington House here. Open every day from 8 pm. Cross the Memorial Bridge from the Lincoln Monument.

7 Washington Monument Ride an elevator to the top of this white obelisk on the Mall, and walk down if you wish. Fine views. Open daily, free.

8 Lincoln Memorial The nation's salute to a martyred president, his monument is one of the most impressive. It's always open.

9 Jefferson Memorial Here the statue of the author of the Declaration of Independence can be seen in a handsome marble rotunda. It's open constantly.

10 The Smithsonian Institution The institution acts as an umbrella organization for a number of Washington's fine museums along the Mall, including the famous National Gallery with its modern extension, the East Building.

11 National Archives Home of the Declaration of Independence, the Bill of Rights and the Constitution. The Building is designed after the Roman Pantheon. Open daily.

12 Hirshhorn Museum and Sculpture Garden This museum was specially designed to display the collection of investor Joseph H. Hirshhorn, who donated it to the nation in 1966. The collection includes more than 200 pieces of sculpture from the 19th century onwards, with over 400 works by 20th-century artists.

13 Georgetown A very charming and popular neighborhood. Georgetown University is located here so there is a large student population. The townhouses are exquisitely elegant. Also home for some of Washington's nightlife.

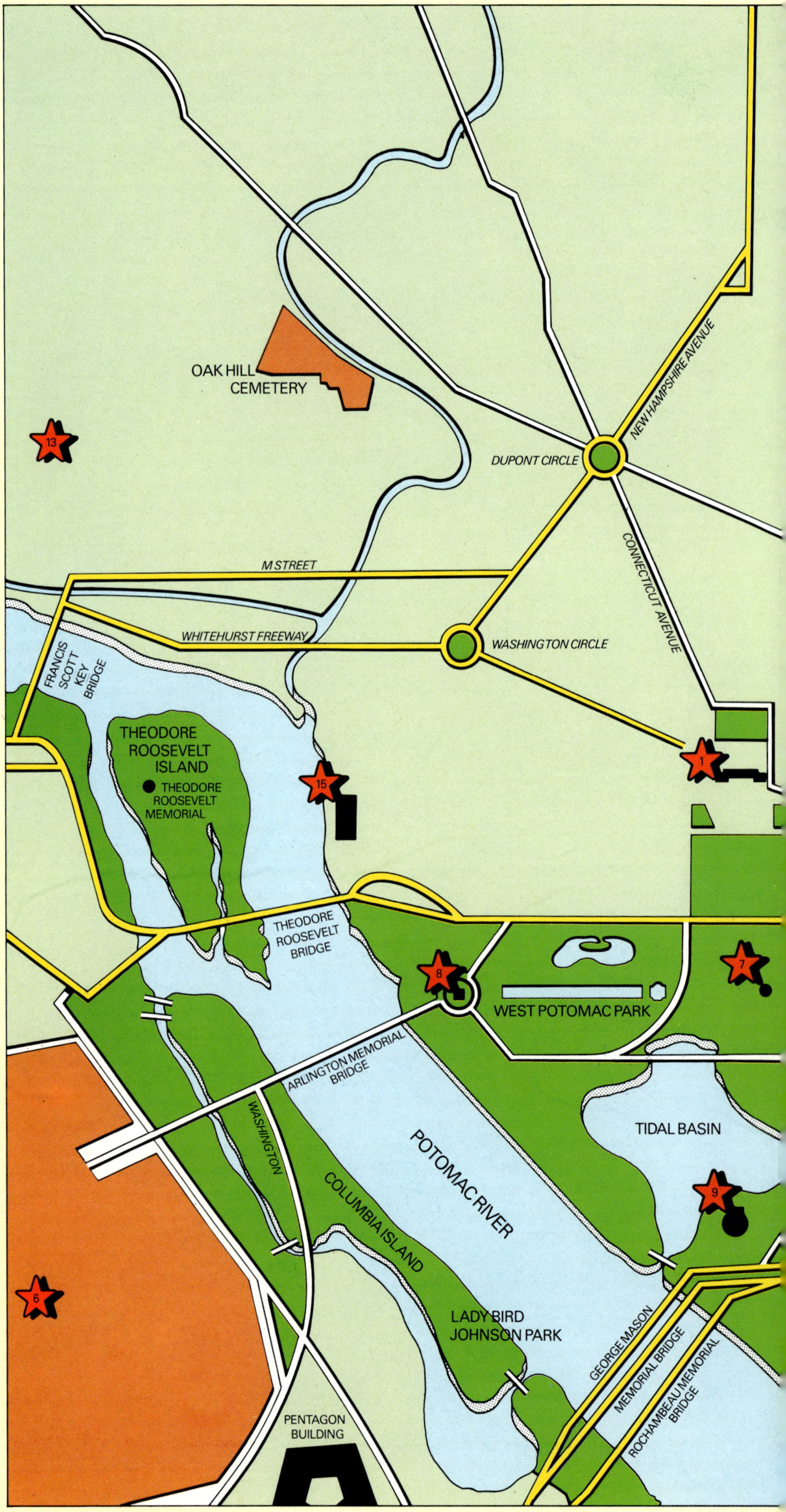

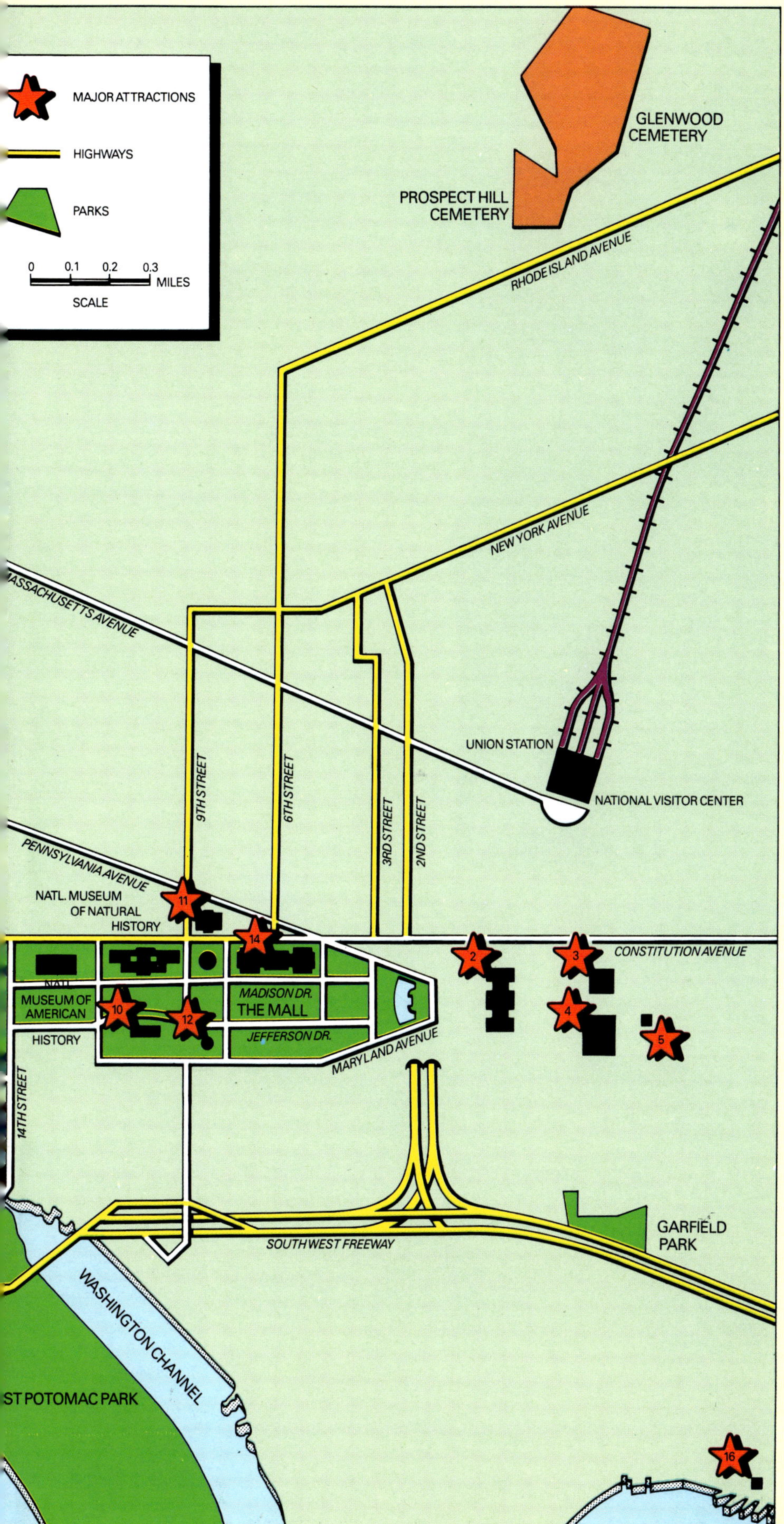

14 National Gallery of Art One of the greatest art museums in the world. Western European and American art from the 13th to the 20th centuries are well-represented.

15 John F. Kennedy Center for the Performing Arts This magnificent building incorporates an opera hall, concert hall and movie theater. It's the official memorial to President Kennedy.

16 U.S. Navy Memorial Museum A great museum for those interested in naval history. This history of the U.S. Navy from 1775 to the space age is charted. Naval weapons are on display and the submarine room has operating periscopes.

Mount Vernon Home of the first President George Washington, just outside the city in Virginia, also on the Potomac. Gardens surround the 18th century House and the graves of George and Martha Washington. (Not shown)

PICTURE CREDITS

Lisa Berg 11 **The Bridgeman Art Library** half title, 14-15, 16 top, 17 **Camerapix Hutchison Library** 8 bottom, 13 bottom, 38 bottom, 53 bottom, 65 left **Bruce Coleman Limited** 11 bottom, 12 top, 28 left, 42, 48 top, 48-49, 50, 52 top, 53 top, 62 **Colorific!** 45 top, 60 top, 63 bottom, David Burnett/Contact Press Images 31 bottom, Paul Conklin 25 top, Mary Fisher 36-37 bottom, Gianfranco Gorgoni/Contact Press Images 31 top, Perorelli/Wheeler Pictures 16 bottom **Daily Telegraph Colour Library** 9 bottom, 13 top, 64 **R. R. Edwards** 40-41, 43 top, 44 top, 58-59 **Angelo Hornak** 18, 72 **The Image Bank** Kay Chernish 54 bottom, C. L. Chryslin 68-69, Patti McConville 23, Bill Thompson 23 bottom **Cynthia Meyer** 10, 43 bottom left, 45 bottom **National Park Services** 49 top, 56-57, 59 top, 65 right, 66 top **Photo Source/Colour Library International** 6-7, 22, 28 right, 32, 34-35, 36-37 top, 46-47, 51, 54 top, 55 **Rex Features** 39, 52 bottom **Frank Spooner Pictures/Gamma** 25 bottom **Tony Stone Photolibrary, London** 24 bottom **USDA** 12 bottom **Vautier-de-Nanxe** front endpaper, 13, 19 top, 29 top, back endpaper **Vision International** contents, 29 bottom, 30, 66 bottom, back cover **WMATA** 33 top left, Larry Levine 33 bottom left, Paul Myatt 8 top, 33 right, 43

Illustration on pages 70-71 by **Mulkern and Rutherford**

MICHIGAN FLORIDA TEXAS IOWA WISCONSIN CALIFORNIA